METAL WORK EDUCATION

Metal work education has long been a critical field in both the industrial and educational sectors, providing essential skills for fabricating, repairing, and innovating with metal-based materials. Metal work, as a vocational trade, encompasses processes like welding, machining, forging, and fabrication

Table of Contents

1. Introduction
1.1 Overview of Metal Work Education
1.2 Importance of Technology in Education
1.3 Aims and Objectives of the Study
2. Historical Background of Metal Work Education
2.1 Evolution of Metal Work from Traditional to Modern Techniques
2.2 Key Developments in Metal Work Education
2.3 Role of Apprenticeship in Early Metal Work Training
3. The Intersection of Technology and Metal Work Education
3.1 Overview of Educational Technology
3.2 Introduction of Computer-Aided Design (CAD) in Metal Work
3.3 CNC Machines and Their Role in Metal Work Training
3.4 Virtual Reality and Simulation in Metal Work Education
4. Impact of Technology on Curriculum Development in Metal Work
4.1 Integration of Technology in Metal Work Curriculum
4.2 The Role of E-Learning Platforms in Metal Work Education
4.3 Challenges and Opportunities in Technology-Enhanced Metal Work Education
5. Technological Tools and Equipment in Metal Work Education
5.1 Computer-Aided Design (CAD) Software
5.2 3D Printing and Metal Fabrication
5.3 Robotics and Automation in Metal Work
5.4 Augmented Reality (AR) and Virtual Reality (VR) Tools
6. Skill Development through Technology
6.1 Role of Simulation in Mastering Metal Work Techniques
6.2 Technology-Enhanced Practical Skills in Metal Work
6.3 The Use of Gamification to Teach Metal Work Concepts
7. The Role of Instructors in Technology-Enhanced Metal Work Education
7.1 Shifting Role of Teachers in the Digital Age
7.2 Professional Development and Training for Metal Work Instructors
7.3 Impact of Technology on Teaching Methodologies
8. Challenges of Integrating Technology in Metal Work Education
8.1 Infrastructure and Cost Limitations
8.2 Resistance to Change in Traditional Educational Models
8.3 Addressing the Digital Divide in Metal Work Education
9. Case Studies of Technology Integration in Metal Work
9.1 Case Study 1: Use of CAD/CAM in Metal Fabrication Programs
9.2 Case Study 2: Robotics in Vocational Metal Work Training

9.3 Case Study 3: Virtual Reality Simulations for Welding Education
10. Global Trends and Future Directions
10.1 Innovations in Metal Work Technology
10.2 Predictions for the Future of Metal Work Education
10.3 Preparing for Technological Disruptions in the Metal Industry
11. Conclusion
11.1 Summary of Findings
11.2 Recommendations for Policy Makers and Educators
11.3 Final Thoughts on the Role of Technology in Metal Work Education

Chapter 1: Introduction

1.1 Overview of Metal Work Education

Metal work education has long been a critical field in both the industrial and educational sectors, providing essential skills for fabricating, repairing, and innovating with metal-based materials. Metal work, as a vocational trade, encompasses processes like welding, machining, forging, and fabrication. This type of education traditionally focused on hands-on skills training, with students learning directly from experienced tradespeople. However, the rapid advancement of technology has begun to revolutionize how metal work is taught and learned.

1.2 Importance of Technology in Education

Over the last few decades, technology has emerged as a transformative force in education. In fields such as engineering, architecture, and manufacturing, the use of technological tools has enhanced both learning efficiency and outcome accuracy. For metal work education, technology not only improves the way learners acquire skills, but it also exposes them to new industry trends and techniques, such as automation, digital fabrication, and computer-aided design. The digital transformation of educational resources has allowed instructors to create dynamic, interactive learning experiences that prepare students for real-world applications.

1.3 Aims and Objectives of the Study

This study aims to explore the impact of technology in enhancing metal work education by:
Analyzing the historical and current role of technology in metal work training.
Examining the types of technological tools and methods applied in metal work education.
Evaluating the effects of technology on teaching methodologies, curriculum design, and student outcomes.
Investigating challenges and barriers to integrating technology in metal work education.
Exploring future trends and innovations that may shape the future of metal work training.

Chapter 2: Historical Background of Metal Work Education

2.1 Evolution of Metal Work from Traditional to Modern Techniques

Metal work has existed for centuries, originating from simple hand-forging techniques to the complex industrial methods seen today. In ancient civilizations, blacksmithing and metal craftsmanship were passed down through apprenticeships, where masters trained students in the essential skills of the trade. This hands-on approach continued through the Industrial Revolution, where technological advancements like steam engines and mechanized tools began to reshape the metal industry. Modern metal work now incorporates a wide range of sophisticated processes, driven by the integration of technology and machinery.

2.2 Key Developments in Metal Work Education

As the metal industry has evolved, so too has metal work education. With the advent of industrial machines and computer technologies, educational institutions have adapted their curricula to include training in digital tools like CNC (Computer Numerical Control) machines and CAD (Computer-Aided Design). These developments have shifted the focus from purely manual techniques to a more holistic approach, where students learn not only traditional skills but also the use of cutting-edge technologies.

2.3 Role of Apprenticeship in Early Metal Work Training

Historically, apprenticeship systems were the cornerstone of metal work education. Students worked under the guidance of a master craftsman, learning by doing, and gradually acquiring the skills necessary to become independent tradespeople. Although this system still exists, it has been supplemented by formal vocational education and training (VET) programs, particularly in the wake of increasing technological complexity in the industry. Today's apprentices are as likely to work with CNC machines or CAD software as they are with hammers and anvils.

Chapter 3: The Intersection of Technology and Metal Work Education

3.1 Overview of Educational Technology

Educational technology encompasses various digital tools and software systems used to improve and streamline the learning experience. In metal work education, this has transformed traditional hands-on training into an integrated learning system, combining both theoretical knowledge and practical applications. Technology such as virtual reality, augmented reality, and machine learning platforms allow students to simulate real-world environments without the physical risks and costs associated with using materials. The growing reliance on digital tools for design, fabrication, and quality control has reshaped the metal work curriculum, pushing educators to include these innovations in their teaching methods.

3.2 Introduction of Computer-Aided Design (CAD) in Metal Work

The integration of CAD software in metal work education has revolutionized the way students approach both design and fabrication. CAD allows students to create precise technical drawings and 3D models, which can be directly used in manufacturing processes such as CNC machining and 3D printing. With the implementation of CAD software, students can learn to:
Visualize complex geometries in three dimensions.
Perform simulations to test the strength, durability, and feasibility of designs.
Make modifications quickly and efficiently before production begins, thus saving on materials and time.
CAD allows students to create precise technical drawings and 3D models, which can be directly used in manufacturing processes such as CNC machining and 3D printing. With the implementation of CAD software, students can learn to:
Visualize complex geometries in three dimensions.

CAD software has become a key component of the metal work curriculum in technical schools and universities worldwide. It also fosters creativity, enabling students to prototype innovative designs that would otherwise be difficult or costly to produce by hand.

3.3 CNC Machines and Their Role in Metal Work Training

Computer Numerical Control (CNC) machines are another technological advancement that has greatly impacted metal work education. These machines automate the manufacturing process, allowing precise control over cutting, shaping, and machining metal parts. CNC technology has been integrated into many educational programs, teaching students how to:
Program and operate CNC machines using G-code or specialized software.
Understand the importance of precision in manufacturing.
Gain hands-on experience in automated production techniques.
The introduction of CNC machines into metal work education provides students with real-world skills that are directly applicable in various industries such as automotive, aerospace, and construction. By learning to work with these machines, students not only enhance their practical skills but also gain an understanding of how automation is changing the landscape of manufacturing.
The introduction of CNC machines into metal work education provides students with real-world skills that are directly applicable in various industries such as automotive, aerospace, and construction. By learning to work with these machines, students not only enhance their practical skills but also gain an understanding of how automation is changing the landscape of manufacturing.

3.4 Virtual Reality and Simulation in Metal Work Education

One of the more recent and cutting-edge technologies being adopted in metal work education is virtual reality (VR) and simulation tools. VR creates immersive environments where students can simulate complex processes such as welding, forging, and assembly in a risk-free, virtual setting. Simulation-based training offers several benefits:
It reduces the cost of materials and consumables, as students can practice without needing physical supplies.
It minimizes safety risks, allowing students to make mistakes in a virtual environment before transitioning to actual metal work.
It provides immediate feedback, helping students to refine their techniques in real-time.
For example, VR welding simulators are now being used in classrooms to teach students the intricacies of welding without the need for expensive equipment and materials. These simulators offer haptic feedback, meaning students can physically feel resistance and heat as they "weld" in the virtual environment, making the learning experience more realistic.

It reduces the cost of materials and consumables, as students can practice without needing physical supplies.

It minimizes safety risks, allowing students to make mistakes in a virtual environment before transitioning to actual metal work.

It provides immediate feedback, helping students to refine their techniques in real-time.

For example, VR welding simulators are now being used in classrooms to teach students the intricacies of welding without the need for expensive equipment and materials.

Chapter 4: Impact of Technology on Curriculum Development in Metal Work

4.1 Integration of Technology in Metal Work Curriculum

The role of technology in metal work education has forced curriculum developers to rethink traditional approaches to training. Today's metal work curriculum must be comprehensive, covering both manual techniques and advanced digital fabrication processes. The integration of technologies like CAD, CNC, and 3D printing has resulted in the following changes:

Theory and Application Balance: Students are no longer only trained in hands-on skills; they also learn to operate complex software and machines that automate these tasks.

Skill Development: The curriculum now includes courses in software proficiency, digital modeling, and automated production systems, along with traditional metal fabrication techniques.

Industry Alignment: The curriculum is being continuously updated to reflect the latest industry standards, ensuring students are prepared for the current job market.

Technology has not only enhanced the depth of metal work training but has also allowed students to learn in more engaging and dynamic ways. Learning Management Systems (LMS) and online platforms offer additional resources, including video tutorials, interactive quizzes, and simulations that can be accessed outside the classroom.

Theory and Application Balance: Students are no longer only trained in hands-on skills; they also learn to operate complex software and machines that automate these tasks.

Skill Development: The curriculum now includes courses in software proficiency, digital modeling, and automated production systems, along with traditional metal fabrication techniques.

Industry Alignment: The curriculum is being continuously updated to reflect the latest industry standards, ensuring students are prepared for the current job market.

4.2 The Role of E-Learning Platforms in Metal Work Education

E-learning platforms have become a vital tool for delivering educational content in many technical disciplines, including metal work. These platforms offer several advantages:

Flexibility: Students can access course materials at their own pace, which is especially important for part-time learners or those balancing education with work commitments.

Access to Resources: E-learning platforms can host a variety of resources, such as video tutorials, downloadable software tools, and instructional guides for using technologies like CAD and CNC machines.

Collaboration: Students and instructors can interact through discussion boards and group projects, fostering collaboration even in distance learning settings.

4.3 Challenges and Opportunities in Technology-Enhanced Metal Work Education

While technology has created numerous opportunities for enhancing metal work education, it also presents certain challenges:

Cost: High-end equipment such as CNC machines, 3D printers, and VR systems are expensive, limiting access for many educational institutions.

Training for Instructors: Educators themselves need to be trained in the use of these advanced technologies, which can require significant time and resources.

Digital Divide: Not all students have equal access to digital tools and resources, particularly in lower-income regions. This creates a challenge in ensuring equitable access to technology-enhanced education.

Nevertheless, the opportunities far outweigh the challenges. Technology has the potential to make metal work education more accessible, more efficient, and more aligned with industry needs.

Training for Instructors: Educators themselves need to be trained in the use of these advanced technologies, which can require significant time and resources.

Digital Divide: Not all students have equal access to digital tools and resources, particularly in lower-income regions. This creates a challenge in ensuring equitable access to technology-enhanced education.

Nevertheless, the opportunities far outweigh the challenges. Technology has the potential to make metal work education more accessible, more efficient, and more aligned with industry needs.

Chapter 5: Technological Tools and Equipment in Metal Work Education

5.1 Computer-Aided Design (CAD) Software

As mentioned in earlier chapters, CAD software is one of the foundational technologies in modern metal work education. Widely used software includes AutoCAD, SolidWorks, and Fusion 360, all of which allow students to:

Design complex metal parts and assemblies.
Perform finite element analysis (FEA) to test structural integrity before production.
Export designs directly to CNC machines or 3D printers for fabrication.
CAD software not only enhances students' understanding of metal work but also prepares them for real-world applications in industries that require precision engineering.

5.2 3D Printing and Metal Fabrication

3D printing, also known as additive manufacturing, has brought new possibilities to metal work education. With 3D printing, students can:
Create detailed metal prototypes layer by layer, reducing material waste compared to traditional methods.
Experiment with complex geometries that would be difficult or impossible to produce using conventional fabrication techniques.
Learn the principles of additive manufacturing, which is becoming increasingly important in industries such as aerospace and medical device manufacturing.
Metal 3D printing technologies, such as Direct Metal Laser Sintering (DMLS), allow students to produce functional metal parts with a high degree of precision, introducing them to the future of metal fabrication.
Experiment with complex geometries that would be difficult or impossible to produce using conventional fabrication techniques.
Learn the principles of additive manufacturing, which is becoming increasingly important in industries such as aerospace and medical device manufacturing.
Metal 3D printing technologies, such as Direct Metal Laser Sintering (DMLS), allow students to produce functional metal parts with a high degree of precision, introducing them to the future of metal fabrication.

5.3 Robotics and Automation in Metal Work

Robotics and automation have become integral to modern manufacturing, and their use in metal work education is growing. By incorporating robotics into the curriculum, students can learn to:
Program industrial robots for tasks such as welding, material handling, and assembly.
Understand how automation improves efficiency and reduces human error in manufacturing.
Gain experience with systems that integrate robotics with other technologies like CNC machining.
Educational programs that include robotics training provide students with a competitive edge in industries where automation is becoming the norm.
Program industrial robots for tasks such as welding, material handling, and assembly.
Understand how automation improves efficiency and reduces human error in manufacturing.

Gain experience with systems that integrate robotics with other technologies like CNC machining.
Educational programs that include robotics training provide students with a competitive edge in industries where automation is becoming the norm.

5.4 Augmented Reality (AR) and Virtual Reality (VR) Tools

In addition to the VR simulations discussed earlier, augmented reality (AR) is another technology being used to enhance metal work education. AR overlays digital information onto the physical world, allowing students to:
Visualize technical drawings and models in 3D directly on top of their work pieces.
Receive step-by-step guidance in real-time, with AR providing visual instructions during the fabrication process.
Collaborate remotely, with instructors able to view and interact with students' work through AR headsets.
AR and VR tools are making it easier for students to grasp complex concepts and improve their practical skills without the risks associated with traditional learning environments.
Visualize technical drawings and models in 3D directly on top of their work pieces.
Receive step-by-step guidance in real-time, with AR providing visual instructions during the fabrication process.
Collaborate remotely, with instructors able to view and interact with students' work through AR headsets.
AR and VR tools are making it easier for students to grasp complex concepts and improve their practical skills without the risks associated with traditional learning environments.

Chapter 6: Skill Development through Technology

6.1 Role of Simulation in Mastering Metal Work Techniques

Simulations are proving to be invaluable in developing key metal work skills. With software such as CNC simulators and virtual welding systems, students can practice repeatedly without the financial and material constraints of real-world applications. This enhances skill acquisition in the following areas:
Precision and accuracy in machine operations.
Safety procedures and protocols.
The ability to troubleshoot and solve technical problems.
Simulations provide a low-risk environment for students to make mistakes, learn, and improve, thus preparing them for real-world scenarios.
Precision and accuracy in machine operations.

Safety procedures and protocols.
The ability to troubleshoot and solve technical problems.
Simulations provide a low-risk environment for students to make mistakes, learn, and improve, thus preparing them for real-world scenarios.

6.2 Technology-Enhanced Practical Skills in Metal Work

Technological tools allow for the development of practical skills that are crucial for the metal work industry. By using CAD software, CNC machines, and robotic arms, students gain hands-on experience in designing, programming, and fabricating metal products. This kind of technology-driven training sharpens their technical proficiency and prepares them for a variety of career paths in sectors such as automotive, aerospace, and heavy machinery manufacturing.
Technological tools allow for the development of practical skills that are crucial for the metal work industry. By using CAD software, CNC machines, and robotic arms, students gain hands-on experience in designing, programming, and fabricating metal products. This kind of technology-driven training sharpens their technical proficiency and prepares them for a variety of career paths in sectors such as automotive, aerospace, and heavy machinery manufacturing.

6.3 The Use of Gamification to Teach Metal Work Concepts

Gamification is another innovative method for skill development in metal work education. By incorporating elements of gaming, such as leaderboards, challenges, and rewards, educators can make learning more engaging.
Gamified platforms help students:
Understand complex concepts in an interactive and enjoyable way.
Stay motivated through competition and rewards.
Receive instant feedback, allowing for continuous improvement.
In the context of metal work education, Gamification can be applied to areas such as:
Virtual welding, where students can compete to achieve the best weld quality.
CNC machining challenges, where students must program machines for optimal efficiency.
Design competitions using CAD software, where students create innovative solutions to real-world problems.
By turning metal work education into an engaging and interactive experience, Gamification encourages students to take ownership of their learning, fostering both theoretical understanding and practical skills.
Understand complex concepts in an interactive and enjoyable way.
Stay motivated through competition and rewards.
Receive instant feedback, allowing for continuous improvement.
In the context of metal work education, Gamification can be applied to areas such as:

Virtual welding, where students can compete to achieve the best weld quality.
CNC machining challenges, where students must program machines for optimal efficiency.
Design competitions using CAD software, where students create innovative solutions to real-world problems.
By turning metal work education into an engaging and interactive experience, Gamification encourages students to take ownership of their learning, fostering both theoretical understanding and practical skills.

Chapter 7: The Role of Instructors in Technology-Enhanced Metal Work Education

7.1 Shifting Role of Teachers in the Digital Age

The advent of technology in metal work education has shifted the role of instructors from being sole knowledge providers to facilitators and mentors. With access to digital resources, students can often learn the basics of techniques through online tutorials, videos, and simulations. As a result, instructors now focus on guiding students through more complex tasks, offering personalized support, and helping students troubleshoot problems as they apply their knowledge in practical settings.

In this new model, instructors are not just teaching specific skills but also helping students develop problem-solving abilities, creativity, and critical thinking. Educators are also required to keep pace with technological advancements, staying updated with the latest tools and methods to ensure that their students are well-prepared for the demands of the industry.

The advent of technology in metal work education has shifted the role of instructors from being sole knowledge providers to facilitators and mentors. With access to digital resources, students can often learn the basics of techniques through online tutorials, videos, and simulations. As a result, instructors now focus on guiding students through more complex tasks, offering personalized support, and helping students troubleshoot problems as they apply their knowledge in practical settings.

7.2 Professional Development and Training for Metal Work Instructors

To fully leverage the potential of technology in metal work education, instructors themselves need continuous professional development. Many educators must learn to use new tools like CNC machines, CAD software, and 3D printers to provide effective instruction. Key areas of training for metal work instructors include:

Technological Proficiency: Learning how to operate and teach with advanced tools and software.

Curriculum Development: Understanding how to integrate these technologies into existing curricula to ensure that students develop both traditional and modern skills.

Pedagogical Skills: Adapting teaching methods to make the most of technology, such as using flipped classrooms, online resources, and blended learning approaches.

Many institutions are investing in professional development programs to help their instructors stay competitive in the rapidly evolving field of technical education.

Technological Proficiency: Learning how to operate and teach with advanced tools and software.

Curriculum Development: Understanding how to integrate these technologies into existing curricula to ensure that students develop both traditional and modern skills.

Pedagogical Skills: Adapting teaching methods to make the most of technology, such as using flipped classrooms, online resources, and blended learning approaches.

Many institutions are investing in professional development programs to help their instructors stay competitive in the rapidly evolving field of technical education.

7.3 Impact of Technology on Teaching Methodologies

Technology has significantly transformed teaching methodologies in metal work education. Instead of focusing solely on lecture-based instruction, educators now employ a range of teaching techniques, including:

Blended Learning: A combination of online and in-person teaching, allowing students to access digital resources outside the classroom while still benefiting from hands-on practice.

Flipped Classrooms: Instructors assign digital lessons (such as video tutorials or simulations) for students to review outside of class. Class time is then dedicated to practical application and problem-solving.

Project-Based Learning: Students work on real-world projects that incorporate the use of advanced technologies like CAD, CNC machines, and 3D printers. This method helps students develop practical skills while engaging with industry-relevant challenges.

These new approaches have increased student engagement and provided more opportunities for individualized learning. Instructors can offer personalized feedback, allowing students to progress at their own pace and focus on areas where they need improvement.

Blended Learning: A combination of online and in-person teaching, allowing students to access digital resources outside the classroom while still benefiting from hands-on practice.

Flipped Classrooms: Instructors assign digital lessons (such as video tutorials or simulations) for students to review outside of class. Class time is then dedicated to practical application and problem-solving.

Project-Based Learning: Students work on real-world projects that incorporate the use of advanced technologies like CAD, CNC machines, and 3D printers. This method helps students develop practical skills while engaging with industry-relevant challenges.

These new approaches have increased student engagement and provided more opportunities for individualized learning. Instructors can offer personalized feedback, allowing students to progress at their own pace and focus on areas where they need improvement.

Chapter 8: Challenges of Integrating Technology in Metal Work Education

8.1 Infrastructure and Cost Limitations

One of the primary challenges of integrating technology into metal work education is the high cost associated with acquiring and maintaining advanced tools and equipment. CNC machines, 3D printers, robotics, and simulation tools require substantial investment, which can be prohibitive for many educational institutions. Additionally, the ongoing cost of software licenses, hardware maintenance, and upgrades can strain budgets.

In regions where funding for vocational education is limited, schools may struggle to provide access to cutting-edge technologies, leading to disparities in the quality of education offered. This can create a divide between institutions that can afford advanced technologies and those that cannot, affecting the employability of students from less well-equipped schools.

In regions where funding for vocational education is limited, schools may struggle to provide access to cutting-edge technologies, leading to disparities in the quality of education offered. This can create a divide between institutions that can afford advanced technologies and those that cannot, affecting the employability of students from less well-equipped schools.

8.2 Resistance to Change in Traditional Educational Models

Resistance to change is another obstacle in integrating technology into metal work education. Many educators and institutions still rely on traditional methods of teaching, focusing primarily on manual techniques and hands-on experience. While these skills remain important, there can be a reluctance to embrace newer technologies such as CAD software, 3D printing, and robotics.

This resistance may stem from a lack of familiarity with digital tools, concerns about the cost and time required for training, or a belief that traditional methods are superior in teaching core metal work skills. Overcoming this resistance requires a shift in mindset, as well as support from administrators to provide the necessary resources and training for educators to become comfortable with technology.

Resistance to change is another obstacle in integrating technology into metal work education. Many educators and institutions still rely on traditional methods of teaching, focusing primarily on manual techniques and hands-on experience. While these skills remain important, there can be a reluctance to embrace newer technologies such as CAD software, 3D printing, and robotics.

8.3 Addressing the Digital Divide in Metal Work Education

The digital divide refers to the gap between those who have access to modern technologies and those who do not. In metal work education, this divide can manifest in several ways:

Access to Equipment: Students in underfunded schools may not have access to advanced tools like CNC machines, CAD software, or 3D printers, putting them at a disadvantage compared to peers in better-funded institutions.

Connectivity: In some regions, poor internet connectivity limits students' ability to access online resources, participate in e-learning, or collaborate on digital platforms.

Skill Gaps: Educators and students in areas with limited technological infrastructure may lack the skills needed to fully engage with advanced technologies, further widening the gap between different schools and regions.

Addressing the digital divide requires investment in infrastructure, funding for schools to acquire the necessary tools, and training programs to ensure that both students and instructors can effectively use new technologies.

Connectivity: In some regions, poor internet connectivity limits students' ability to access online resources, participate in e-learning, or collaborate on digital platforms.

Skill Gaps: Educators and students in areas with limited technological infrastructure may lack the skills needed to fully engage with advanced technologies, further widening the gap between different schools and regions.

Addressing the digital divide requires investment in infrastructure, funding for schools to acquire the necessary tools, and training programs to ensure that both students and instructors can effectively use new technologies.

Chapter 9: Case Studies of Technology Integration in Metal Work

9.1 Case Study 1: Use of CAD/CAM in Metal Fabrication Programs

This case study examines the use of Computer-Aided Design (CAD) and Computer-Aided Manufacturing (CAM) software in metal fabrication programs. In many technical schools, CAD/CAM integration has allowed students to design metal components on a computer, simulate their performance, and then produce them using CNC machines.

One example is a technical college in the United States that introduced CAD/CAM as a core part of its metal work curriculum. Students learned to design parts using SolidWorks and AutoCAD, and then used CAM software to create tool paths for CNC machines. The program reported an increase in student engagement and improved job placement rates, as graduates were better prepared for the demands of modern manufacturing jobs.

This case highlights the importance of digital design and manufacturing tools in preparing students for careers in industries that require precision and automation.

This case study examines the use of Computer-Aided Design (CAD) and Computer-Aided Manufacturing (CAM) software in metal fabrication programs. In many technical schools, CAD/CAM integration has allowed students to design metal components on a computer, simulate their performance, and then produce those using CNC machines.

One example is a technical college in the United States that introduced CAD/CAM as a core part of its metal work curriculum. Students learned to design parts using SolidWorks and AutoCAD, and then used CAM software to create tool paths for CNC machines. The program reported an increase in student engagement and improved job placement rates, as graduates were better prepared for the demands of modern manufacturing jobs.

This case highlights the importance of digital design and manufacturing tools in preparing students for careers in industries that require precision and automation.

9.2 Case Study 2: Robotics in Vocational Metal Work Training

In this case study, a vocational school in Germany integrated robotics into its metal work training program. The school partnered with local industry leaders to provide students with access to industrial robots used in manufacturing processes such as welding, material handling, and assembly.

Students were taught how to program and operate robots, learning skills that are in high demand in industries such as automotive and aerospace manufacturing. The school reported that students who completed the robotics program had higher employability rates and were more competitive in the job market.

This case study demonstrates how integrating robotics into metal work education can help bridge the gap between education and industry, providing students with the skills they need to succeed in an increasingly automated world.

Students were taught how to program and operate robots, learning skills that are in high demand in industries such as automotive and aerospace manufacturing. The school reported that students who completed the robotics program had higher employability rates and were more competitive in the job market.

This case study demonstrates how integrating robotics into metal work education can help bridge the gap between education and industry, providing students with the skills they need to succeed in an increasingly automated world.

9.3 Case Study 3: Virtual Reality Simulations for Welding Education

A technical institute in Canada implemented virtual reality (VR) welding simulators to enhance its welding education program. The simulators provided students with an immersive learning experience, allowing them to practice welding techniques in a virtual environment before working with actual materials.

The institute found that students who used the VR simulators were able to master basic welding skills more quickly than those who relied solely on traditional methods. The simulators also reduced material costs and minimized safety risks, as students could practice in a controlled environment without the need for consumables like metal or gas.

This case study highlights the potential of VR technology to improve the efficiency and effectiveness of metal work education, particularly in areas where safety and cost are major concerns.

A technical institute in Canada implemented virtual reality (VR) welding simulators to enhance its welding education program. The simulators provided students with an immersive learning experience, allowing them to practice welding techniques in a virtual environment before working with actual materials.

The institute found that students who used the VR simulators were able to master basic welding skills more quickly than those who relied solely on traditional methods. The simulators also reduced material costs and minimized safety risks, as students could practice in a controlled environment without the need for consumables like metal or gas.

This case study highlights the potential of VR technology to improve the efficiency and effectiveness of metal work education, particularly in areas where safety and cost are major concerns.

Chapter 10: Global Trends and Future Directions

10.1 Innovations in Metal Work Technology

The field of metal work is continuously evolving, with new technologies and processes emerging that promise to further transform the industry. Key innovations include:

Additive Manufacturing: Metal 3D printing is advancing rapidly, with new techniques such as binder jetting and directed energy deposition offering faster and more cost-effective ways to produce metal parts.

AI and Machine Learning: Artificial intelligence and machine learning are being used to optimize metal fabrication processes, from predictive maintenance of equipment to real-time quality control.

Advanced Materials: New metal alloys and composites are being developed that offer superior strength, durability, and corrosion resistance, opening up new possibilities for metal work applications.

These innovations are likely to shape the future of metal work education, as students will need to learn new skills and techniques to work with these advanced technologies.

10.2 Predictions for the Future of Metal Work Education

As technology continues to advance, the future of metal work education is likely to be characterized by:

Increased Automation: With more manufacturing processes becoming automated, students will need to focus on learning how to program and maintain automated systems, rather than relying solely on manual skills.

Personalized Learning: Adaptive learning technologies could allow students to receive personalized instruction based on their individual strengths and weaknesses, leading to more efficient and effective learning outcomes.

Global Collaboration: As the world becomes more interconnected, metal work students may have the opportunity to collaborate with peers and industry professionals from around the globe, using digital platforms to work on joint projects and share knowledge.

Increased Automation: With more manufacturing processes becoming automated, students will need to focus on learning how to program and maintain automated systems, rather than relying solely on manual skills.

Personalized Learning: Adaptive learning technologies could allow students to receive personalized instruction based on their individual strengths and weaknesses, leading to more efficient and effective learning outcomes.

Global Collaboration: As the world becomes more interconnected, metal work students may have the opportunity to collaborate with peers and industry professionals from around the globe, using digital platforms to work on joint projects and share knowledge.

10.3 Preparing for Technological Disruptions in the Metal Industry

The metal industry is poised for significant disruption as new technologies such as artificial intelligence (AI), robotics, and additive manufacturing continue to reshape the landscape. To prepare for these technological disruptions, both educational institutions and industry professionals need to embrace flexibility and continuous learning. The following strategies can help educators and students remain competitive:

- Embracing Lifelong Learning: As technologies evolve rapidly, it is essential for metal work professionals to adopt a mindset of lifelong learning. Educational institutions should offer continuing education and certification programs that allow workers to update their skills in areas like CNC programming, robotics, and AI-enhanced manufacturing processes.
- Partnerships with Industry: Strong partnerships between educational institutions and industry leaders will be crucial in keeping curricula relevant and up-to-date. By working closely with businesses, schools can ensure that students are learning the latest technologies and techniques being used in the field.
- Interdisciplinary Education: The metal work industry is becoming increasingly interdisciplinary, with overlaps in fields such as engineering, computer science, and

materials science. Metal work education programs should integrate aspects of these fields to provide students with a broader skill set, making them more adaptable in a changing industry.
- ➢ Preparing for Sustainable Manufacturing: As environmental concerns take center stage, the future of metal work will also be influenced by the demand for sustainable practices. This includes the development of energy-efficient manufacturing processes and the use of recyclable or eco-friendly materials. Educational programs must incorporate sustainability into their curricula to prepare students for future challenges in the industry.
- ➢ By taking proactive measures, the metal work education sector can ensure that students are equipped with the knowledge and skills needed to thrive in an industry that is continually being transformed by technological innovations.

Chapter 11: Conclusion

11.1 Summary of Findings

This extensive study explored the transformative role of technology in metal work education, highlighting how digital tools, automated machinery, and cutting-edge innovations have redefined both the teaching and learning experience. Key takeaways from the research include: Technology, such as CAD, CNC, robotics, and virtual reality, has significantly enhanced metal work education by providing more efficient, safe, and cost-effective training environments. The integration of these tools into the curriculum allows students to develop both traditional manual skills and advanced technical knowledge, ensuring they are well-prepared for the demands of modern industry.

While challenges such as cost, resistance to change, and the digital divide remain, the opportunities presented by technology are vast, offering a more engaging, accessible, and forward-looking approach to metal work education.

This extensive study explored the transformative role of technology in metal work education, highlighting how digital tools, automated machinery, and cutting-edge innovations have redefined both the teaching and learning experience. Key takeaways from the research include: Technology, such as CAD, CNC, robotics, and virtual reality, has significantly enhanced metal work education by providing more efficient, safe, and cost-effective training environments. The integration of these tools into the curriculum allows students to develop both traditional manual skills and advanced technical knowledge, ensuring they are well-prepared for the demands of modern industry.

While challenges such as cost, resistance to change, and the digital divide remain, the opportunities presented by technology are vast, offering a more engaging, accessible, and forward-looking approach to metal work education.

11.2 Recommendations for Policy Makers and Educators

To maximize the benefits of technology in metal work education, several recommendations emerge from this study:

Investment in Infrastructure: Governments and educational institutions must prioritize funding for the acquisition of advanced equipment such as CNC machines, CAD software, 3D printers, and virtual reality tools. Investing in technology infrastructure is critical to ensuring that students receive hands-on experience with industry-relevant tools.

Teacher Training and Professional Development: Continuous professional development for educators is essential. Instructors should receive ongoing training in the use of emerging technologies to effectively incorporate them into their teaching methodologies.

Industry Collaboration: Schools should foster stronger partnerships with industries to align educational programs with the latest technologies and trends. Such collaboration ensures that students graduate with skills that are directly applicable in the workforce.

Inclusive Access to Technology: Policymakers should work toward closing the digital divide by ensuring that all students, regardless of their geographic location or socioeconomic status, have access to the necessary technology to succeed in metal work education. This could involve subsidies for schools in underserved areas and programs that provide students with laptops, software, and internet access.

Teacher Training and Professional Development: Continuous professional development for educators is essential. Instructors should receive ongoing training in the use of emerging technologies to effectively incorporate them into their teaching methodologies.

Industry Collaboration: Schools should foster stronger partnerships with industries to align educational programs with the latest technologies and trends. Such collaboration ensures that students graduate with skills that are directly applicable in the workforce.

Inclusive Access to Technology: Policymakers should work toward closing the digital divide by ensuring that all students, regardless of their geographic location or socioeconomic status, have access to the necessary technology to succeed in metal work education. This could involve subsidies for schools in underserved areas and programs that provide students with laptops, software, and internet access.

11.3 Final Thoughts on the Role of Technology in Metal Work Education

As metal work education continues to evolve, the integration of technology will play an increasingly important role in shaping the future of the industry. Technology has the power to enhance traditional learning, foster innovation, and prepare students for careers in a rapidly changing world. By embracing the latest advancements and adopting flexible, forward-thinking educational models, institutions can ensure that metal work education remains relevant, effective, and accessible for future generations.

Extended Bibliography and Resources

To support the development of this comprehensive study, the following resources can be utilized:

Books on the history and evolution of metal work and industrial design.

Academic journals focusing on vocational and technical education, with an emphasis on the impact of technology.

Industry white papers and reports on the future of manufacturing, including trends in automation, AI, and additive manufacturing.

Case studies and technical manuals on the use of CAD, CAM, CNC, and virtual reality in educational settings.

Technology has the power to enhance traditional learning, foster innovation, and prepare students for careers in a rapidly changing world. By embracing the latest advancements and adopting flexible, forward-thinking educational models, institutions can ensure that metal work education remains relevant, effective, and accessible for future generations.

Extended Bibliography and Resources

To support the development of this comprehensive study, the following resources can be utilized:

Books on the history and evolution of metal work and industrial design.

Academic journals focusing on vocational and technical education, with an emphasis on the impact of technology.

Industry white papers and reports on the future of manufacturing, including trends in automation, AI, and additive manufacturing.

Reference

Groover, M. P. (2020). Automation, Production Systems, and Computer-Integrated Manufacturing. Pearson.
Kalpakjian, S., & Schmid, S. R. (2016). Manufacturing Engineering and Technology. Pearson.
Chua, C. K., Leong, K. F., & Lim, C. S. (2010). Rapid Prototyping: Principles and Applications. World Scientific Publishing.
Ashby, M. F., Shercliff, H., & Cebon, D. (2013). Materials: Engineering, Science, Processing and Design. Butterworth-Heinemann.
Zhang, D. (2020). Metal Additive Manufacturing: Materials, Design, Technologies, and Applications. Springer.
Oladejo, M. A., & Olaniyan, O. (2020). "Effectiveness of Computer-Aided Design in the Teaching and Learning of Metal Work in Technical Colleges." Journal of Technical Education and Training, 12(2), 123-135.
Cohen, A., & Jeong, H. (2019). "Using Virtual Reality for Training in the Metal Fabrication Industry." Journal of Vocational Education & Training, 71(4), 486-500.
Goodson, A., & McCabe, M. (2021). "Integrating Additive Manufacturing into Metal Work Curricula: Challenges and Benefits." International Journal of Engineering Education, 37(1), 78-87.
Rogers, M., & Holloway, S. (2018). "The Use of Robotics in Technical and Vocational Education: A Case Study of Automation in Metal Work Training." Journal of Technical and Vocational Education Research, 14(3), 245-260.
Dagli, C. (2017). "Computer Numerical Control (CNC) Machines in Vocational Education: A Global Perspective." International Journal of Technical Education, 9(2), 65-82.
Zhou, H., & Wang, Y. (2018). "The Role of Technology in Enhancing Metal Work Education: Case Studies from China." Proceedings of the International Conference on Vocational Education and Training (ICVET), pp. 201-215.
Pena, L. M., & Andrews, T. J. (2019). "Virtual Reality and Simulation-Based Training in Welding Education." Proceedings of the American Society for Engineering Education (ASEE), pp. 432-445.
Smith, R. A., & Murphy, C. (2021). "Enhancing Metal Work Education Through Gamification: A New Approach." Proceedings of the European Conference on Educational Research (ECER), pp. 158-172.
World Economic Forum (2020). The Future of Jobs Report 2020: Focus on Metal Work and Manufacturing.
Manufacturing Institute (2019). Advanced Manufacturing: Closing the Skills Gap in the Metal Industry.
National Center for Education Statistics (2018). Technology in Vocational Education
Autodesk Education - https://www.autodesk.com/education
American Welding Society (AWS) - https://www.aws.org/education

Fab Foundation - https://fabfoundation.org

6. Theses and Dissertations

Kumar, R. (2017). "Exploring the Role of Advanced Technologies in Metal Work Vocational Education." PhD dissertation, University of Michigan.

This dissertation examines the integration of advanced technologies such as CNC machines, 3D printing, and robotics into vocational metal work programs, with a focus on their educational and economic impacts.

Davies, J. (2020). "Virtual Reality Simulations in Metal Work Education: A Case Study of Welding Training Programs." Master's thesis, University of Edinburgh.

Almeida, F. (2018). "The Influence of Gamification on Student Engagement in Technical Metal Work Courses." PhD dissertation, University of São Paulo.

Krar, S., Gill, A., & Smid, P. (2010). Technology of Machine Tools. McGraw-Hill Education.

Rehg, J. A., & Kraebber, H. W. (2021). Computer-Integrated Manufacturing. Pearson.

Seidel, R., & Pernul, G. (Eds.). (2015). Advances in Information Systems for Metal Work Education. Springer.

Tlusty, G. (1999). Manufacturing Processes and Equipment. Prentice Hall.

Zhang, X., & Gao, S. (2020). "Exploring the Role of Additive Manufacturing in Metal Work Education." Journal of Vocational and Technical Education, 45(3), 221-240.

Johnston, B., & Ryan, G. (2018). "Computer-Aided Design (CAD) in Technical Metal Work Education: A Review of Pedagogical Approaches." Journal of Technical and Engineering Education, 14(4), 76-92.

Chen, J., & Peng, W. (2019). "The Influence of CNC Programming on Metal Work Students' Cognitive Skills." International Journal of Engineering Education, 35(2), 312-324.

Williams, T., & Becker, H. (2020). "Evaluating the Use of Robotics in Metal Fabrication Education." Journal of Advanced Manufacturing Education, 12(2), 189-203.

Nguyen, H. L., & Tran, P. T. (2021). "The Use of Augmented Reality (AR) in Metal Work Technical Training: A Comparative Study." Journal of Educational Technology and Society, 24(1), 45-59.

WorldSkills International (2020). Skills for the Future: The Role of Technology in Vocational Education.

National Association of Manufacturers (NAM) (2019). Digital Transformation in Manufacturing Education: Closing the Skills Gap.

Deloitte (2021). Advanced Manufacturing Workforce Study: Future of Metal Work and Fabrication Jobs.

European Union Agency for Vocational Education (2019). Adapting Vocational Education to the Digital Age: Metal Work and Beyond.

The Fabricator - https://www.thefabricator.com

Machinery's Handbook Online - https://www.machineryhandbook.com

Welding Digest by AWS - https://weldingdigest.aws.org

Fusion 360 for Education - https://www.autodesk.com/products/fusion-360/education
Martinez, A. (2019). "Exploring the Impact of Digital Fabrication Technologies in Metal Work Education." PhD dissertation, Stanford University.
Walker, D. (2020). "A Study on the Integration of CNC Machining into Metal Work Curricula in Community Colleges." Master's thesis, University of Southern California.
Li, Q. (2021). "The Effectiveness of Virtual Reality Welding Simulators in Technical Education." PhD dissertation, University of British Columbia.
Taylor, R. (2018). "Teaching Automation in Metal Fabrication: A New Pedagogical Approach." PhD dissertation, University of Sheffield.

www.ingramcontent.com/pod-product-compliance
Lightning Source LLC
Chambersburg PA
CBHW071002220526
45471CB00007B/3140